Mnemonics
in
BIOCHEMISTRY

Mnemonics *in* BIOCHEMISTRY

Wilma Delphine Silvia CR MBBS MD DNB MNAMS
Professor and Head
Department of Biochemistry
Akash Institute of Medical Sciences and Research Centre
Bengaluru, Karnataka, India

Ravish H MBBS MD
Assistant Professor
Department of Biochemistry
Sapthagiri Institute of Medical Sciences and Research Centre
Bengaluru, Karnataka, India

The Health Sciences Publishers
New Delhi | London | Philadelphia | Panama

Jaypee Brothers Medical Publishers (P) Ltd

Headquarters

Jaypee Brothers Medical Publishers (P) Ltd
4838/24, Ansari Road, Daryaganj
New Delhi 110 002, India
Phone: +91-11-43574357
Fax: +91-11-43574314
Email: jaypee@jaypeebrothers.com

Overseas Offices

J.P. Medical Ltd
83 Victoria Street, London
SW1H 0HW (UK)
Phone: +44 20 3170 8910
Fax: +44 (0)20 3008 6180
Email: info@jpmedpub.com

Jaypee Medical Inc
The Bourse
111 South Independence Mall East
Suite 835, Philadelphia, PA 19106, USA
Phone: +1 267-519-9789
Email: jpmed.us@gmail.com

Jaypee Brothers Medical Publishers (P) Ltd
Bhotahity, Kathmandu, Nepal
Phone: +977-9741283608
Email: kathmandu@jaypeebrothers.com

Jaypee-Highlights Medical Publishers Inc
City of Knowledge, Bld. 237, Clayton
Panama City, Panama
Phone: +1 507-301-0496
Fax: +1 507-301-0499
Email: cservice@jphmedical.com

Jaypee Brothers Medical Publishers (P) Ltd
17/1-B Babar Road, Block-B, Shaymali
Mohammadpur, Dhaka-1207
Bangladesh
Mobile: +08801912003485
Email: jaypeedhaka@gmail.com

Website: www.jaypeebrothers.com
Website: www.jaypeedigital.com

Inquiries for bulk sales may be solicited at: jaypee@jaypeebrothers.com

Mnemonics in Biochemistry

First Edition: **2014**, Reprint: 2024

ISBN 978-93-5152-392-5

Printed in India

Dedicated to

All the students who have inspired us
to write this book

Preface

The basic essence of medicine is biochemistry. Students find it very difficult to recall the subject from their memory. Each student will be following different routes of learning. Memory needs every method of nurturing its capacity. Mnemonic is a learning technique that aids information retention. Mnemonic systems are special techniques or strategies consciously used to improve memory. It also helps to employ information already stored in long-term memory to make memorization an easier task. Mnemonics included in this book have been collected from various sources as well as synthesized by the authors. We thank our students of Sapthagiri Institute of Medical Sciences and Research Centre, who have been a source of continued inspiration in preparing this mnemonics book.

Wilma Delphine Silvia CR
Ravish H

Contents

Chemistry of Amino Acids and Proteins

Cell

Enzymes

Biological Oxidation

Vitamins

Carbohydrate Metabolism

Amino Acid Metabolism

Lipid Metabolism

Heme Metabolism

Acid-Base Balance

Mineral Metabolism

Nutrition

Detoxification

Nucleic Acids Chemistry and Metabolism

Molecular Biology

Hormones

Others

CHEMISTRY OF AMINO ACIDS AND PROTEINS

1. Essential Amino Acids

H Histidine
T Tryptophan
T Threonine
P Phenylalanine
V Valine
I Isoleucine
L Leucine
L Lysine
M Methionine
A Arginine

2. Semi-essential Amino Acids

H Histidine
A Arginine

3. Non-essential Amino Acids

C Cysteine
A Aspartate
A Asparagine

T Tyrosine
S Serine

G Glycine
G Glutamate
G Glutamine
A Alanine
P Proline

Or

A Alanine
C Cysteine
T Tyrosine

G Glycine
A Asparagine
S Serine
P Proline

G Glutamate
A Aspartate
G Glutamine

4. Structural Classification of Amino Acids

"Attractive Handsome Actor Amitabh Bachchan's Studio Image"	**A**cidic **H**ydroxyl **A**liphatic **A**romatic **B**asic **S**ulfhydryl **I**mino

5. Diamino Acids

"Hindustan Aeronautics Limited"	**H**istidine **A**rginine **L**ysine

6. Types of Bonds in Protein Structure

"PhD"	**P**eptide bonds—Primary **H**ydrogen bonds—Secondary **D**isulfide bonds—Tertiary

7. Branched Chain Amino Acids

"LeVIs"	**Le**ucine **V**aline **Is**oleucine

8. Hydrophobic Amino Acids

"VIMAL tried Proline Pants"	**V**aline **I**soleucine **M**ethionine **A**lanine **L**eucine **T**ryptophan **P**roline **P**henylalanine

9. Amino Acids with Aliphatic Side Chains

"GALIV"/ "Going Airtel Live Is Valid"	**G**lycine **A**lanine **L**eucine **I**soleucine **V**aline

10. Amino Acids in a Chain of Insulin

In Hindi EKKIS = In English 21	A Chain = **21**AA

11. Basic Amino Acids

"HAL"	**H**istidine **A**rginine **L**ysine

CELL

12. Golgi Complex: Functions

"Golgi—Dispatches A SPAM"	**D**istributes proteins and lipids from endoplasmic reticulum **A**dd mannose onto specific lysosome proteins **S**ulfation of sugars and selected tyrosine **P**roteoglycan assembly

	Add O-oligosugars to serine and threonine **M**odify N-oligosugars on asparagine

13. Metabolic Pathways in Cytosol

FGGH–"Foreign Graduate Girls Home"/ "Fancy Girls Going Home"	**F**A synthesis **G**lycogen synthesis **G**lycolysis **H**MP shunt

ENZYMES

14. Enzymes: Competitive Inhibitors

"Competition is hard because we have to travel more kilometers (km) with the same velocity"	With competitive inhibitors, velocity remains same but K_m increases

15. Enzymes: Classification

"Over The Hill Lives Isolated Lady"	**O**xidoreductases **T**ransferases

Hydrolases
Lyases
Isomerases
Ligases

BIOLOGICAL OXIDATION

16. Electron Transport Chain, Complex I Inhibition

"Roten-One"	**R**otenone is a site specific inhibitor of complex I

17. Inhibitor of Electron Transport Chain

"CO blocks CO"	Carbon monoxide (**CO**) blocks **C**ytochrome **O**xidase (CO)

VITAMINS

18. Vitamin K-dependent Substances

"Naughty Students Tend To Sleep Cozily"	**N**ine **S**even

Ten
Two
Protein **S**
Protein **C**

19. Fat Soluble Vitamins

"Eat Daily Kashmir Apple"	Vitamin **E** Vitamin **D** Vitamin **K** Vitamin **A**

20. Water Soluble Vitamins

"The Rhythm Neatly and Properly Played Continuously"	**T**hiamine **R**iboflavin **N**iacin **P**antothenic acid **P**yridoxine **C**obalamin

21. Hypervitaminosis A: Signs and Symptoms

"Increased Vitamin A makes you"	**H**eadache **H**epatomegaly

"Happy, Helpful, Attractive, Admiring, Daring and Dashing Producer Boney"	**A**norexia, **A**lopecia, **D**ry skin, **D**rowsiness and **P**ainful **B**ones

22. Pellagra Manifestation

	Dermatitis **D**iarrhea **D**ementia

23. Antioxidant Vitamins

"ACE"	Vitamin **A** Vitamin **C** Vitamin **E**

24. Vitamin C: Functions

"Cute Curious Talented Children Cherished Television Program In International School For Blind"	**C**ataract prevention **C**hronic diseases prevention **T**yrosine metabolism **C**ollagen formation **C**orticosteroid synthesis **T**ryptophan metabolism

Peptide hormone synthesis
Immunological
Iron metabolism
Sparing action of other vitamins
Folic acid metabolism
Bone formation

25. Vitamin K-dependent Blood Clotting Factors

"Star News To Telecast"	**S**even **N**ine **T**en **T**wo

26. Thiamine Pyrophosphate-dependent Enzymes

"Bear Attacked The Popeye"	**B**ranched chain alpha-keto acid dehydrogenase **A**lpha-ketoglutarate dehydrogenase **T**ransketolase **P**yruvate dehydrogenase

27. Pyridoxine (Vitamin B_6): Functions

"Gangam's Special Talented Thrilling And Dashing Disco Dance Night Show"	**G**lycogen phosphorylase activity **S**erine synthesis **T**ransamination-schiff base **T**ranssulfuration **A**mino acids absorption **D**ecarboxylation **D**eamination **D**elta-aminolevulinic acid synthesis **N**iacin coenzymes **S**tones—hyperoxaluria

28. Biotin: Functions

"Glamorous The Fatty Princess Loved"	**G**luconeogenesis **T**CA cycle **F**atty acid synthesis **P**ropionyl-CoA **L**eucine metabolism

29. Causes of Vitamin B_{12} Deficiency

"Humpty Dumpty Got Auto Home"	**H**ereditary malabsorption of vitamin
	Dietary deficiency
	Gastrectomy
	Autoimmune destruction of gastric mucosa
	HCl insufficiency

30. Water-soluble Vitamin-like Compounds

"Petrol Bunk Located In Chikkasandra"	**P**ABA
	Bioflavonoids
	Lipoic acid
	Inositol
	Choline

CARBOHYDRATE METABOLISM

31. Citric Acid Cycle

"Anyway Our College Is Always Safe and Sound From Mosquitoes"

Acetyl CoA
Oxaloacetate
Citrate
Isocitrate
Alpha-ketoglutarate
Succinyl-CoA
Succinate
Fumarate
Malate

Or

"Anyway Oxalo Citrate Is Of A Silly Suspense: Fumbling Molecule"

Acetyl CoA
Oxaloacetate
Citrate
Isocitrate
Oxalosuccinate
Alpha-ketoglutarate
Succinyl-CoA
Succinate
Fumarate
Malate

32. Enzymes of Tricarboxylic Acid Cycle

"City Accounts Inspectors Always Spoke Slander For Money"	**C**itrate synthase **A**conitase **I**socitrate dehydrogenase **A**lpha-ketoglutarate dehydrogenase **S**uccinate thiokinase **S**uccinate dehydrogenase **F**umarase **M**alate dehydrogenase

33. Three Irreversible Reactions in Glycolysis

"Pappa Pray God"	**P**yruvate kinase **P**hosphofructokinase **G**lucokinase

34. Glycogen Storage Disorders

"Very Pampered Child Admitted to Medical Hospital Temporarily"/	**v**on Gierke's disease **P**ompe's disease **C**ori's disease

"Volatile Pills Cause A Major Hulk Temporarily"	**A**nderson's disease **M**cArdle's disease **H**er's disease **T**arui's disease

35. Deficient Enzyme in Glycogen Storage Disorders

"A-B"	Andersen's—**B**ranching enzyme
"C-D"	Cori's—**D**ebranching enzyme
"M-M"	McArdle's—**M**uscle glycogen phosphorylase
"H-H"	Her's—**H**epatic glycogen phosphorylase

36. Enzymes of Glycolysis

"High Profile People Act Perfectly Great, Posing Priority Every Places"	**H**exokinase **P**hosphoglucoisomerase **P**hosphofructokinase **A**ldolase **P**hosphotriose isomerase **G**lyceraldehyde-3-phosphate dehydrogenase **P**hosphoglycerate kinase

Phosphoglycerate mutase

Enolase

Pyruvate kinase

37. Gluconeogenesis

"Pathway Produces Fresh Glucose"	**P**yruvate carboxylase **PEP** carboxykinase **F**ructose-1,6-bisphosphatase **G**lucose-6-phosphatase

38. G6PD Deficiency

"5H's"	**H**MP shunt enzyme deficiency **H**emolysis of RBCs by exposing them to oxidation (which normally is prevented by NADPH) **H**emoglobin is altered **H**einz bodies formation **H**emolytic anemia

39. Glycolysis

"Goodness Gracious, Father Francis Did Go By Picking Plums (to) Prepare Pastry"	**G**lucose **G**lucose-6-phosphate **F**ructose-6-phosphate **F**ructose-1,6-bisphosphate **D**ihydroxyacetone-phosphate **G**lyceraldehyde-3-phosphate 1,3-**B**isphosphoglycerate 3-**P**hosphoglycerate 2-**P**hosphoglycerate **P**hosphoenolpyruvate **P**yruvate

40. G6PD: Oxidant Drug-inducing Hemolytic Anemia

"AAA-3A's"	**A**ntibiotic (e.g. sulfamethoxazole) **A**ntimalarial (e.g. primaquine)

	Antipyretics (e.g. acetanilide, but not aspirin or acetaminophen)

41. Galactosemia: Enzyme Deficiency

"GALIPUT"	**Gal**actose-1-**p**hosphate **u**ridyl**t**ransferase

42. Glucagon: Function

"Mr Gluca has Gone by cycling to AMP to bring out some Glucose"	Glucagon elevates glucose by cAMP mechanism

43. Pyruvate: Products of Complete Oxidation

"4 Navigators + 1 Farmer + 3 Cops + 1 Girl"	Complete oxidation of pyruvate yields: **4 N**ADH + **1 F**ADH$_2$ + **3** CO$_2$ + **1 G**TP

44. Diabetic Ketoacidosis

"DKA signs DKA"	**D**ehydrated **K**etones/Kussmaul breathing/K$^+$ drops **A**cidosis

45. Malate-aspartate Shuttle

"MAD commute"	**M**alate in **A**lpha-ketoglutarate and **D** (aspartate) out

46. Intermediates of Krebs Cycle or TCA Cycle

"Abdul Abraham Gentlemen Got Presidentship"	**A**spartic acid **A**sparagine **G**lutamic acid **G**lutamine **P**roline

AMINO ACID METABOLISM

47. BUN Elevation—Causes

"ABCD"	**A**zotemia (pre-renal) **B**leeding (GI) **C**atabolic status **D**iet (high protein parenteral nutrition)

48. Initial Three Steps of Catabolism of Branched-chain Amino Acids

"Tamed Ox Dehydrated" Or *"TOD"*	**T**ransamination **O**xidative decarboxylation **D**ehydrogenation

49. Salient Features of Transamination Reactions

"Partying Special Ninjas Required Sound N Dance"	**P**LP **S**pecific transaminase **N**o ammonia **R**eversible **S**ynthesis of non-essential amino acids **N**o energy **D**iagnostic

50. Maple Syrup Urine Disease

"Isolated Lucy Validated Oxford"	**I**soleucine **L**eucine **V**aline **O**xidative decarboxylation defect

51. Amino Acids Not Taking Part in Transamination Reactions

"Lousy Teachers Penalize Heroism"	**L**ysine **T**hreonine **P**roline **H**ydroxyproline

52. Disposal of Ammonia

"Americans Urinate Urea"	**A**mmonotelic **U**ricotelic **U**reotelic

53. Metabolic Defects in Urea Cycle

"American Army Officers Are Capable"	**A**rgininosuccinase **A**rginase **O**rnithine transcarbamoylase **A**rgininosuccinate synthetase **C**arbamoyl phosphate synthetase I

54. Specialized Products from Glycine

"George Capable of Hacking Powerful Computers"

Glutathione
Creatine
Heme
Purine
Conjugated bile acids

55. Specialized Products from Tyrosine

"Talented Teachers Educate Needy Doped Males"

Thyroxine
Triiodothyronine
Epinephrine
Norepinephrine
Dopamine
Melanin

56. Specialized Products from Tryptophan

"Serve Neat Meat"

Serotonin
NAD$^+$/NADP$^+$
Melatonin

57. Specialized Products from Methionine

"Sam Clearly Purchase at E-bay"

SAM
Creatinine
Polyamines
Epinephrine

58. Specialized Products from Cysteine

"Cops Target Active Gluttons Constantly"	**C**ysteine **T**aurine **A**ctive sulfate **G**lutathione **C**oenzyme A

59. Specialized Products from Arginine

"No Crops"	**N**itric oxide **C**reatine

60. Specialized Products from Lysine

"Librarian's Cards"	**L**ysine → **C**arnitine

61. Specialized Products from Glutamate

"Glutton and Gambist Gabbar"	**G**lutathione **G**amma-carboxyglutamate **G**amma-aminobutyric acid

62. Specialized Products from Glutamine

"Amina's Pure Pyramid"	**A**mino sugars **P**urine **P**yrimidine

63. Specialized Products from Aspartate

"Pure Piracy"	**P**urine **P**yrimidine

64. Specialized Products from Serine

"Specific Phospho Cholinergic"	**S**phingomyelin **P**hosphatidylserine **C**holine

65. Beta-alanine Product

"Bear's cannibalize"	**B**eta-alanine → CoA

66. Methylated Products

"Many Modern Marines Crossed Egypt Cities Perfectly And Making Peace"	**M**etanephrine **M**ethylated RNA bases **M**ethylnicotinamide **C**holine **E**pinephrine **C**reatine **P**hosphatidylcholine **A**nserine **M**ethylated amino acids **P**rotein

67. Methyl Group Acceptor

"Nice Elephants Entered Pond Playfully Grazing Neatly All Bamboo Stems Cleverly"	**N**icotinamide **E**pinephrine **E**thanolamine **P**hosphatidylethanolamine **P**rotein (amino acid) **G**uanidinoacetate **N**orepinephrine **A**cetyl serotonin **B**ases **S**erine **C**arnosine

68. One-carbon Metabolism

"Heavy Money Makes Man's Future Fine"	**H**ydroxymethyl **M**ethyl **M**ethylene **M**ethenyl **F**ormyl **F**ormimino

69. Polyamines

"Prudent Sperms"	**P**utrescine **S**permine

70. Alkaptonuria

"Black Beauty Horse Obey Coach"	**B**lack urine disease **B**enedict's test **H**omogentisate oxidase **O**chronosis **C**oke in color

71. Tyrosinosis

"Free Man Roamed rejoiced Peacefully Carefully Righteous Life"	**F**umarylacetoacetate hydroxylase **M**aleylacetoacetate isomerase **R**enal tubular dysfunction **P**olyneuropathy **C**abbage-like odor **R**ickets **L**iver failure

72. Albinism

"Men Race Cars Teasing Police Man"	**M**elanosomes **R**ecessive **C**ancer

Tyrosinase deficiency
Photophobia
Melanin

73. Tryptophan Metabolism—Kynurenine Pathway

"Tasting Nicely Fried Noodles, King and Queen Had Horrible Pain"	**T**ryptophan **N**AD$^+$ **F**ormylkynurenine **N**icotinate **K**ynurenine 2-amino-3-carboxymuconate semialdehyde **Q**uinolate **H**ydroxykynurenine **H**ydroxyanthranilate **P**icolinate

74. Metabolism of Tryptophan—Serotonin and Melatonin Synthesis

"Terrorist Having Shotgun Hijacked All Males"	**T**ryptophan **H**ydroxytryptophan **S**erotonin

	Hydroxyindoleacetate **A**cetylserotonin **M**elatonin

75. S-adenosylmethionine Cycle

"Master Sachin's High Score"	**M**ethionine **S**-adenosylmethionine **H**omocysteine **S**-adenosylhomocysteine

76. Synthesis of Cysteine from Methionine

"Many Hazardous Carcinogens Always Cause Sickness"	**M**ethionine **H**omocysteine **C**ystathionine **A**lpha-ketobutyrate **C**ysteine **S**uccinyl-CoA

77. Metabolism of Cysteine and Cystine

"Cockroaches Carefully Move Crevices, Pick Catastrophic Things Storing Stinky Parasitic Manure"	**C**ystine **C**ysteine **M**ercaptoethanolamine **C**ysteine sulfinate **P**yruvate

Cysteic acid
Taurine
Sulfite
Sulfate
PAPS
Mucopolysaccharides

78. Branched Chain Amino Acid Metabolism

"Kangaroos Always Got Freedom"	**K**eto acids **A**cyl-CoA thioesters **G**lucose **F**at

79. Glutamate Products

"Arrogant Gorillas Never Gain Peace"	**A**mmonia **G**lutathione **N**-acetylglutamate **G**amma-carboxyglutamate **G**lucose **P**roteins

80. Overview of Glutamine Metabolism

"Pink Panther Angrily Defeated Powerful Army"	**P**urines
	Pyrimidines
	Ammonia
	Detoxification
	Proteins
	Amino sugars

81. Steps in Metabolism of Gamma Aminobutyric Acid

"God Gila Sucked Successfully"	**G**lutamate
	GABA
	Succinate semialdehyde
	Succinate

82. Aspartate Metabolism

"People Purchase Always Good Projectors Urgently"	**P**yrimidines
	Purines
	Asparagine
	Glucose
	Proteins
	Urea

83. Overview of Serine Metabolism

"Sinbad Sailor Prevented Plans of Alibaba Crossing Gujarat Gates Easily"

Sphingomyelin
Selenocysteine
Proteins
Phosphatidylserine
One-carbon metabolism
Alanine
Cysteine
Glycine
Glucose
Ethanolamine

84. Overview of Threonine Metabolism

"German Police Arrested Large Public"

Glycine
Proteins
Alpha-ketobutyrate
Lactate
Pyruvate

85. Glucogenic and Ketogenic Amino Acids

"Lazy Prison Inmates Terrorized Truly"/

Lysine
Phenylalanine
Isoleucine
Tyrosine
Tryptophan

"Tyson Traveled Lovely Philippine Islands"	**T**yrosine **T**ryptophan **L**ysine **P**henylalanine **I**soleucine

86. Sources of Carbon Skeleton for Synthesis of Non-essential Amino Acids

"Sweetly → Cuddle Gorillas"	**S**erine → **C**ysteine and **G**lycine
"Pretty → Alan"	**P**yruvate → **A**lanine
"Police → Siren"	3-**P**hosphoglycerate → **S**erine
"Physical → Training"	**P**henylalanine → **T**yrosine
"Intermediates TCA Cycle → GAP"	**I**ntermediates of **TCA** **C**ycle → **G**lutamic acid, **A**spartic acid and **P**roline

87. Disorders of Phenylalanine and Tyrosine

"Popular Televisions Need Attractive and Talented Actors"	**P**henylketonuria **T**yrosinemia type II **N**eonatal tyrosinemia

	Alkaptonuria **T**yrosinosis/ **T**yrosinemia type I **A**lbinism

88. Disorders of Sulfur Containing Amino Acids

"Ceylon Communist Hijacked I, II, III, IV Cylinders"/ *"3 Cy(si)sters, Homes I, II, III, IV"*	**Cy**stinuria **Cy**stinosis **Cy**stathioniuria **Ho**mocystinuria type I **Ho**mocystinuria type II **Ho**mocystinuria type III **Ho**mocystinuria type IV

89. Cystinuria Caused due to Defect in Renal Tubular Amino Acid Transporter for

"COLA"	**C**ystine **O**rnithine **L**ysine **A**rginine

90. Urea Cycle

"Careless Citizens Argued And Fumbled About Urination Obviously"	**C**arbamoyl phosphate **C**itrulline **A**rgininosuccinate **A**spartate **F**umarate **A**rginine **U**rea **O**rnithine

LIPID METABOLISM

91. Fatty Acid Synthesis and Oxidation

"SIT and Synthesize"	CITRATE → synthesis of fatty acid CARnitine → **B**urning **F**at

92. Major Apolipoproteins

"ABCE"	**A**-1 activates lecithin-cholesterol acyltransferase (LCAT) **B**-100 binds to low density lipoprotein (LDL) receptor

C-II cofactor for lipoprotein lipase
E-mediates extra (remnant) uptake

93. Sphingolipidoses

"Naughty Fat Godzilla Kindly Thanked Father"	**N**iemann-Pick disease **F**arber's disease **G**aucher's disease **K**rabbe's disease **T**ay-Sachs disease **F**abry's disease

94. Enzyme Defect in Sphingolipidoses

"Students Always Bunked Biochemistry Classes Happily"	**S**phingomyelinase **A**lpha-galactosidase **B**eta-glucosidase **B**eta-galactosidase **C**erebrosidase **H**exosaminidase

95. Five Stages of Cholesterol Synthesis

"Sun Flower Production for Successful Company"	**S**ynthesis of HMG-CoA **F**ormation of mevalonate (6C) **P**roduction of isoprenoid units (5C) **S**ynthesis of squalene (30C) **C**onversion of squalene to cholesterol

96. Cholesterol Synthesis

"Adventurous Accidents Hurt Marching Public People, Isolated Dancing Girls For Squashing Large Community"	**A**cetyl-CoA **A**cetoacetyl-CoA **H**MG-CoA **M**evalonate **P**hosphomevalonate **P**yrophosphomevalonate **I**sopentenyl pyrophosphate **D**imethylallyl pyrophosphate **G**eranyl pyrophosphate

Farnesyl pyrophosphate
Squalene
Lanosterol
Cholesterol

97. Steroid Hormone Synthesis

"Co-operative Pregnant Progressively Curtailed Alcohol Easily"	**C**holesterol **P**regnenolone **P**rogesterone **C**ortisol **A**ldosterone **E**stradiol

98. Management of Hypercholesterolemia

"Perfect Lifestyle Always Favor People"	**P**UFA **L**ifestyle modifications **A**void cholesterol/ **A**void carbohydrate diet **F**ibers **P**lant sterol

99. Classification of Lipoproteins

"Confused Villain Intentionally Loved His Friends"	**C**hylomicrons **V**ery low density lipoprotein (VLDL) **I**ntermediate density lipoprotein (IDL) **L**ow density lipoprotein (LDL) **H**igh density lipoprotein (HDL) **F**ree fatty acids

100. Fabry's Disease

"FABRY'S"	**F**oam cells in glomeruli/**F**ebrile episodes **A**lpha-galactosidase A deficiency/ **A**ngiokeratomas **B**urning pain in extremities, **B**oys, **B**UN increased in serum **R**enal failure

YX phenotype (male, X linked recessive)
Sphingolipidoses

101. Prostaglandin Functions

"Use Brain Passionately Keep Simple Peaceful Lifestyle"	**U**terus **B**rain **P**latelet aggregation **K**idney functions **S**ecretions **P**ain and fever **L**ocal hormone

102. Ketone Bodies Seen in

"Type I DM vs Type II DM"	ket**ONE** bodies are seen in type **ONE** diabetes

103. HDL vs LDL

H is Healthy L is Lethal	HDL—**H** is **H**ealthy or **H**ighly desirable LDL—**L** is **L**ethal

HEME METABOLISM

104. Metabolic Pathways in Both Cytoplasm and Mitochondria

"HUG"	**H**eme synthesis **U**rea cycle **G**luconeogenesis

105. Porphyrias

"All Congenital Porphyrias Have Variable Presentation"	**A**cute intermittent porphyria **C**ongenital erythropoietic porphyria **P**orphyria cutanea tarda **H**ereditary coproporphyria **V**ariegate porphyria **P**rotoporphyria

106. van den Bergh Reaction

"I"ndirect reacting bilirubin = "U"nconjugated bilirubin	Both start with vowels, so they go together

107. Oxygen Dissociation Curve: Causes of Shift to Right

"CAT BE right!"	**C**O_2 **A**cid **T**emperature 2,3- **B**isphosphoglycerate **E**xercise
Or	
"TEA BisCuits"	**T**emperature **E**xercise **A**cid 2,3- **B**isphosphoglycerate **C**O_2

108. Conjugated and Unconjugated Hyperbilirubinemia

C: "DR"	**C**onjugated hyperbilirubinemia: **D**ubin-Johnson syndrome **R**otor syndrome

U: "Creative Gilbert"	**U**nconjugated hyperbilirubinemia: **C**rigler-Najjar syndrome **G**ilbert's syndrome

109. Substrates of Heme Synthesis

"ALA Placed in Urine CUP Produces Prototype Heme"	**A**minolevulinic acid **P**orphobilinogen **U**roporphyrinogen **C**oproporphyrinogen **P**rotoporphyrinogen **P**rotoporphyrin **H**eme

110. Heme Catabolism

"2HB Uro Media Stereo"	**H**emoglobin **H**eme **B**ilirubin **B**iliverdin **U**robilinogen **M**esobilinogen **S**tercobilinogen

111. Bilirubin: Phototherapy

"BiLirUbin → BLU'"	**B**i**L**ir**U**bin absorbs light maximally in the **BLU**e range

112. Bilirubin: Common Causes for Increased Levels

"OH Liver"	**O**bstruction **H**emolysis **L**iver disease

ACID-BASE BALANCE

113. Normal Anion Gap Metabolic Acidosis (NAGMA): Causes

"Doctor's Children Happy"	**D**iarrhea **C**arbonic anhydrase inhibitor **H**yperkalemia

114. Metabolic Acidosis: Causes

"USED CAR"	**U**remia **S**alicylate poisoning **E**thylene glycol poisoning

	Diabetic ketoacidosis **C**arbonic anhydrase inhibitors **A**mmonium chloride **R**enal tubular acidosis

115. Metabolic Changes in Alkalosis

Think: "Al-K-loss and Al-Ca-loss"	Loss of **K**$^{+}$ causing hypokalemia
Thus, in a state of alkalosis, there is	Loss of **Ca**$^{++}$ causing hypocalcemia

116. Anion Charge

"AN → A N"	**AN**ion is **A** **N**egative ion

117. Anion Gap Acidosis: Pathophysiology-based Approach

"KULT"	**K**etones: Diabetic, alcoholic and starvation ketoacidosis Mechanism: Acetoacetate, beta-hydroxybutyrate

Uremia: Renal failure (usually creatinine > 5 mg/dL)
Mechanism: Phosphates, sulfates, other organic acids
Lactic acidosis: Sepsis
Toxins: Toxic alcohols, aspirin
Mechanism: Formic acid and oxalic acid; salicylic acid

118. Alkalosis vs Acidosis: Directions of pH and HCO_3^-

"ROME"

Respiratory = **O**pposite:

- pH is high, pCO_2 is down (alkalosis)
- pH is low, pCO_2 is up (acidosis)

Metabolic = **E**qual:

- pH is high, HCO_3 is high (alkalosis)
- pH is low, HCO_3 is low (acidosis)

119. Acids: Lewis Acid vs Bronsted Acid

"BAD LATE"

BAD: **B**ronsted **A**cid **D**onates hydrogens
LATE: **L**ewis **A**cid **T**akes **E**lectrons

MINERAL METABOLISM

120. Zinc-dependent Enzymes

"A CLAS"

Alcohol dehydrogenase
Carbonic anhydrase
Lactate dehydrogenase
Alkaline phosphatase
Superoxide dismutase

121. Essential Trace Elements

"In China, I Met Zero Mileage Car From Sunny Chennai"

Iron
Copper
Iodine
Molybdenum
Zinc
Manganese
Cobalt

Fluorine
Selenium
Chromium

Or

"Many Cute Children Celebrated Innocent Father Zee's Mangalore fancy store Inauguration"	**M**olybdenum **C**hromium **C**opper **C**obalt **I**ron **F**luorine **Z**inc **S**elenium **M**anganese **I**odine

122. Functions of Calcium

"Many Secretaries Calmly Messaged Bloody Secrets Contacting Bonny, Never Enemies Heard"	**M**embrane integrity **S**ecretion processes **C**almodulin **M**uscle contraction **B**lood **S**econdary messenger **C**ontact inhibition

Bones
Nerve transmission
Enzymes
Heart

123. Hypercalcemia: Causes

"PM DIPS ME"	**P**arathyroid (hyperparathyroidism) **M**alignancy **D**iuretics (thiazide the main culprit) **I**mmobilization/ Idiopathic **P**aget's disease **S**arcoidosis **M**egadoses of vitamins A and D **E**ndocrine (Addison's disease, thyrotoxicosis)

124. Functions of Phosphorous

"Popular Bollywood Actor Gagan Begged New Producer"	**P**roteins activation **B**uffer **A**TP

GTP
Bone
NAD$^+$
Phospholipids

125. Functions of Copper

"Malgudi's Homeless Elephant Hungrily Chased Many Children"

Myelin
Heme
Enzymes
Hemocyanin
Ceruloplasmin
Melanin
Collagen

126. Functions of Zinc

"Empty Gardens Allowed Indian Venomous Cobras"

Enzymes
Gustin
Antioxidant
Insulin
Vitamin A
Cell growth

127. Functions of Selenium

"The 21st Gentleman Very Heroically Married To Lovely Cinderella"	**T**hioredoxin **21**st amino acid **G**lutathione **V**itamin E action **H**eavy metals chelation **M**embrane integrity **T**hyroxin to triiodothyronine **L**ipid peroxidation prevention **C**ancer prevention

128. Functions of Iron

"PIC Transport"	**P**eroxidase **I**mmunocompetence **C**ytochromes **T**ransport of oxygen and carbon dioxide

NUTRITION

129. Kwashiorkor

"MEAL"	**M**alabsorption **E**dema **A**nemia **L**iver (fatty)

DETOXIFICATION

130. Salient Features of Cytochrome P450

"Hungry Lion Picked Innocent Nervous Antilopes"	**H**emoproteins **L**iver (most) **P**hospholipids **I**nducible enzymes **N**ADPH **A**romatic hydrocarbons

131. Conjugation Reactions

"Glamorous Gorgeous Gentle Guy Showed Assets to Sultan"	**G**lucuronic acid **G**lycine **G**lutathione **G**lutamine

S-Adenosyl methionine
Acetic acid
Thiosulfate
Sulfate

132. Utilization of Energy in Human

"Bahujan Samaj Party"	**B**asal metabolic rate **S**pecific dynamic action **P**hysical activity

133. Factors Affecting Basal Metabolic Rate

"Sweet Sexy Angelina Played Harmoniously Enjoying Delicious Food @ Saturday's Reception"	**S**urface area **S**ex **A**ge **P**hysical activity **H**ormones **E**nvironment **D**isease states **F**ever **S**tarvation **R**ace

NUCLEIC ACIDS CHEMISTRY AND METABOLISM

134. Pyrimidines

"CUT"	**C**ytosine **U**racil **T**hymine

135. Composition of Nucleotides and Nucleosides

Nucleotide has BSP: "Bahujan Samaj Party"	Composition of Nucleotides: **B**ases, **S**ugar (ribose) and **P**hosphates
Nucleoside has BS: "Blood Sugar"	Composition of Nucleosides: **B**ases and **S**ugar (ribose)

136. Lesch-Nyhan Syndrome

"Self Confident Men Get Heavy Success"	**S**elf-mutilation **C**horeoathetosis **M**ental retardation **G**out **H**yperuricemia **S**pasticity

137. Pseudogout Crystals

"2P's"	**P**olygon shaped **P**ositive birefringent

138. Initiating Codon

"AUG: Augments"	**Augments** (initiates) the protein synthesis **AUG** also codes for methionine

139. Blotting Techniques

"South Indian dish: Dosa	**S**outhern blotting for **D**NA
North Indian dish: Roti	**N**orthern blotting for **R**NA
West Indian dish: Paratha"	**W**estern blotting for **P**roteins

140. Stop Codons

"U Go Away	**UGA**
U Are Away	**UAA**
U Are Gone"	**UAG**

MOLECULAR BIOLOGY

141. Single Base Alteration

"Dead Ducks Dangerously Incorporated Insects"	**D**eamination **D**epurination **D**eletion **I**ncorporation of base analog, alkylation **I**nsertion

142. Repair of DNA

"Basic Nuclear Missiles Doubled"	**B**ase excision repair **N**ucleotide excision repair

Or

"Busy Narendra Miserably Danced"	**M**ismatch repair **D**ouble strand break repair

143. Transcription by Prokaryotes

"Pretty 35 Red Roses"	**P**ribnow box —thirty five **(35)** **R**NA polymerase **R**ho dependent and Rho independent

144. Transcription by Eukaryotes

"Hogness Enjoyed Tasting Cat Porridge"	**H**ogness box **E**nhancer **T**ranscription factor **C**AAT box **P**olymerases-RNA

145. Exon vs Intron Function

"EX → EX, IN → IN"	**Ex**ons **Ex**pressed **InTr**ons **In Tr**ash

146. Inhibitors of Transcription

"Ali's Active Riffles"	**A**lpha-amanitin **A**ctinomycin D **R**ifampin

147. Sickle Cell Anemia: Mutation

"Sixth Hemoglobin isn't Very Good"	At **S**ixth position of **H**b beta chain, **V**aline is present instead of **G**lutamic acid

148. Total RNA

"Homeless, Restless, Soundless, Senseless, Tasteless Male Serpents"	**h**nRNA **r**RNA **s**nRNA **s**noRNA **t**RNA **m**RNA **s**cRNA

149. Characteristics of Genetic Code

"Waiter Served Ultimate Delicious Noodles"	**W**obble hypothesis **S**pecificity **U**niversality **D**egenerate **N**on-overlapping

150. Alternate Names for Recombinant DNA Technology

"New Robots Cloned Magnificent Genes"	**N**ew genetics **R**ecombinant **C**loning **M**anipulation of genes **G**enetic modifications

151. Vectors—the Cloning Vehicles

"Bactericidal Artificial Plastic Cosmetics"	**B**acteriophages **A**rtificial chromosome as vectors **P**lasmids **C**osmids

152. Basic Techniques in Genetic Engineering

"Innovative Blogs Sequentially Transferred Power Monopolized Construction Site"	**I**solation of nucleic acids **B**lotting techniques **S**equencing-DNA **T**ransfer of genes **P**olymerase chain reaction **M**onoclonal antibodies **C**onstruction of gene library **S**ite-directed mutagenesis

HORMONES

153. Pituitary Hormones

"GOAT FLAP"	**G**rowth hormone **O**xytocin **A**drenocorticotropin **T**hyroid-stimulating hormone **F**ollicle-stimulating hormone **L**uteinizing hormone **A**ntidiruetic hormone (vasopressin) **P**rolactin

154. Insulin Resistance: Features

"UPHOLD Me Father"	Elevated plasma **U**ric acid Increased **P**lasminogen activated inhibitor-1 **H**yperinsulinemia and hypertension

Obesity (central)
Low HDL
Diabetes mellitus (type 2)
Microalbuminuria
Increased **F**ibrinogen

155. One of the Functions of Insulin

"PG"

INsulIN stimulates 2 things to go IN 2 cells:
Potassium
Glucose

156. Goiter: Causes

"GOITER"

Goitrogens
Onset of puberty
Iodine deficiency
Thyrotoxicosis/Tumor/Thyroiditis (Hashimoto's)
Enzyme deficiencies
Reproduction (pregnancy)

157. Aldosterone: Regulation of Secretion from Adrenal Cortex

"RANS"	**R**enin-angiotensin mechanism **A**trial natriuretic peptide (ANP) **N**a concentration in blood **S**tress

158. Adrenaline Mechanism

"ABC of Adrenaline"	**A**drenaline → activates **B**eta receptors → increases **C**yclic AMP

159. Adrenal Cortex Hormones

"Salt-Sugar and Stress-Sex"	Aldosterone—**sodium** balance (zona glomerulosa) Glucocorticoids—**sugar** from glucose Major **stress** hormone (zona fasciculata)

Androgens—**sex** hormones (zona reticularis)

OTHERS

160. Collagen Concisely Covered

"COLLAGEN"	**C**-terminal propeptide (procollagen)/ Covalent cross-links/C vitamin/Connective tissue/Cartilage/ Chondroblasts/Copper cofactor (covalent cross-linking)
	Outside the cell is where collagen normally functions/ Osteoblasts/ Osteogenesis imperfecta

Lysyl hydroxylase/Lysyl oxidase (oxidatively deaminates lysyl and hydroxylysyl residues to form collagen cross-links, last biosynthesis step)

Long triple helical fibers/Ligaments

Alpha chains/Attached by H bonds form triple helix/Ascorbate for hydroxylation of lysyl and prolyl residues of pro-alpha chains (post-translational modification)

Glycine in every third position/Glycosylation of hydroxyl group of hydroxylysine with Glucose and Galactose; GOlgi allows procollagen to GO outside of cell

Extracellular matrix/
Eye (cornea, sclera)/
Ehlers-Danlos syndrome
N-terminal propeptide (procollagen)/Non-helical terminal extensions

161. Collagen Sites

"Be Calm Read Books"	**B**one **C**artilage **R**eticulin **B**asement membrane

162. X-linked Recessive Disorders

"Hari Got A Dull Life Made Miserable With Retaining Females Fabrication"	**H**emophilia **G**6PD deficiency **A**gammaglobulinemia **D**iabetes insipidus **L**esch-Nyhan syndrome **M**enke's disease **M**uscle dystrophy **W**iskott-Aldrich syndrome

Retinitis pigmentosa
Feminization syndrome
Fabry's disease

163. Functions of Glutathione

"TMEDICS"

Transport of amino acids
Methemoglobin reduction
Erythrocyte homeostasis (inactivation of free radicals inside RBC)
Detoxification of organophosphorus, halogenated and nitrogenous compounds
Insulin inactivation
Coenzyme in reduction reactions
Scavenger of toxic substances

164. Nitric Oxide Synthase (NOS): Types

"MEN"	**M**acrophage NOS (inducible NOS) **E**ndothelial NOS **N**euronal NOS

165. Gibbs Free Energy Formula

"Good Honey Tastes Sweet"	$\Delta \mathbf{G} = \mathbf{H} - \mathbf{T}\Delta \mathbf{S}$

166. Oxidation

"OIL"	**O**xidation **I**s **L**oss (of electrons)

167. Reduction

"RIG"	**R**eduction **I**s **G**ain (of electrons)

168. Hyponatremia: Causes

"A DEAD HORSE"	**A**typical pneumonia **D**rugs **E**ndocrine (hypothyroidism)

Addison's disease
Diuretics
Hypovolemia
Oedema (edema)
Renal
Syndrome of inappropriate antidiuretic hormone secretion (SIADH)
Exsanguination

169. Hypernatremia: Causes

"6 D's"

Dehydration
Diarrhea
Diuretics
Diabetes insipidus
Doctors (iatrogenic)
Disease: kidney, sickle cell, etc.

170. Hypokalemia: Causes

"BAD LOAD"

Bartter/Conn's syndrome (hyperaldosteronism)

Alkalosis
Diuretics
Laxative abuse
Other causes: insulin overdose
Acute glucose load
Diarrhea

171. Cell Cycle Stages

"Go Sweety Go! Make Candles!"	**G**1 phase (growth phase 1) **S** phase (DNA synthesis) **G**2 phase (growth phase 2) **M** phase (mitosis) **C** phase (cytokinesis)

172. Cell Division

"Professor Metal Announced Telephonically"	**P**rophase **M**etaphase **A**naphese **T**elophase

173. Calculi: Types

"CAlCULi"	**C**alcium **A**mmonium magnesium phosphate **C**ystine **U**ric acid

174. Anemia: Total Iron Binding Capacity (TIBC) Finding to Differentiate Iron Deficiency vs Chronic Disease

"TIBC"	**TIBC** levels at the **T**op = **I**ron deficiency **B**ottom = **C**hronic disease

175. Alpha Fetoprotein (AFP): Causes for Increased Maternal Serum AFP During Pregnancy

"Increased Maternal Serum Alpha-FetoProtein"	**I**ntestinal obstruction **M**ultiple gestation/ **M**iscalculation of gestational age/ **M**yeloschisis **S**pina bifida cystica

Anencephaly/
Abdominal wall defect
Fetal death
Placental abruption

176. Increased Alpha Fetoprotein in Blood is seen in

"TOLD"	**T**esticular tumors **O**bituary—fetal death **L**iver—hepatomas **D**efects—neural tube defects

177. Sequence of Elevation of Enzymes after Myocardial Infarction

"Tropical C-AST-Le"	**T**roponin **C**K-MB **AST** **L**DH1

178. Protein Content of Milk

"Humming CowBoys Go in multiples of 1.1"	**H**uman: 1.1 **C**ow: 2.2 **B**uffalo: 3.3 **G**oat: 4.4

179. Transferrin: Normal Levels

Normal level of transferrin: "Trans-Ferrin"	The normal level of transferrin in blood is 2–5 g/L (Two-Five)
Normal range of Transferrin Saturation: "Trans-Ferrin Saturation"	The normal transferrin saturation is 20%–60% (Twenty-Sixty) (The value is 25%–56% to be more accurate)

180. Excitatory Neurotransmitters

"Pure SAND GlAss"	**P**urines **S**erotonin **A**cetyl choline (ACh) **N**oradrenaline/ adrenaline **D**opamine **Gl**utamic acid **A**spartic acid